-8 APR 1986

The local search series
Editor: Mrs Molly Harrison M B E
Formerly Curator of the Geffrye Museum, London E2

Living Creatures of an English Home

A timber-framed house. Can you see signs of creatures living in it?

Living Creatures of an English Home

Olive Royston BA

Line drawings by Elizabeth Clarke

London   Routledge & Kegan Paul

First published in 1971 by Routledge & Kegan Paul Ltd
Broadway House, 68-74 Carter Lane, London, EC4V 5EL
Photoset and printed in Great Britain by
BAS Printers Ltd, Wallop, Hampshire
© Olive Royston 1971
No part of this book may be reproduced in any form without
permission from the publisher except for the quotation of brief
passages in criticism
ISBN 0 7100 6919 7 (C)
ISBN 0 7100 6920 0 (P)

The local search series
Editor: Mrs Molly Harrison MBE
Formerly Curator of the Geffrye Museum, London, E2

Many boys and girls enjoy doing research about special topics and adding drawings, photographs, tape-recordings and other kinds of evidence to the notes they make. We all learn best when we are doing things ourselves.
The books in this series are planned to help in this kind of 'project' work. They give basic information but also encourage the reader to find out other things; they answer some questions but ask many more; they suggest interesting things to do, interesting places to visit, and other books that can help readers to enjoy their finding out and to look more clearly at the world around them.

M.H.

'. . . all the business of life is to endeavour to find out what you don't know by what you do'
John Whiting *Marching Song*

Contents

		page
	Editor's preface	viii
1	Your project	1
2	Houses and their human inhabitants	7
3	Pets and other mammals	13
4	Birds	20
5	Insects	30
6	Spiders and mites	41
7	Visitors	48
8	Parasites	59
	Conclusion – points for your project	67
	Some more things to do	68
	How to keep living creatures	69
	Acknowledgments	70

Editor's preface

It may seem strange to you to have a book about 'creatures' in a house. Houses are, of course, built for people, so we tend to think that only people should live in them, and that a house with other living things in it must be a dirty house.

I think this book will help you to see things differently, to enjoy finding out who else lives in your home besides you and your family, and to think about why they are there.

Many people nowadays are beginning to realize that modern men and women, boys and girls, are very destructive, even when they do not mean to be. Because there are many of us, and because we want space for living, for working, for travelling about and for our leisure, we are all the time killing trees, plants, birds, fish, animals and insects. Unless we think carefully about what we do, we destroy many of them more quickly than they can reproduce themselves, and this can be dangerous for us all. If we 'upset the balance of nature' in this way, we may cause serious problems for the future, so it is worth discussing and thinking about all this, and helping in any 'conservation' work if we can.

The creatures written about in this book are small ones, many of them are harmless and many of them, you will find, share your home with you.

If you study them, draw them and make notes about them, you will begin to realize that they are as important in their way, as you are in yours. You will probably begin to like them, too, and if you have to move them, you will want to do so gently, to a place where they can be as comfortable as they have been in your home.

M.H.

Your project   1

Living creatures in a house form a number of 'communities' linked together

How to set about your project on this subject — how to make it interesting — how to plan your work — how to find the facts

Most human beings live in families and so do other smaller creatures. Whether or not a particular kind of creature will be able to live in any one special place — a wood, a meadow, a pond or a house, for example — will depend upon what other kinds of animals, and of plants, live there. This is because all the plants and animals living in one place depend on each other and so form what we call a *community*.
Nowadays, too, most people live in some sort of community, such as a village or a town. This means that some members of the community produce goods — food, clothing and much else. Others provide services — shops, transport, schools. What others can you think of?
We all rely on each other. In fact it is practically impossible today for a man in any place in the world to live his life independently of other people. Even adventurers like Sir Francis Chichester or the Russian and American astronauts live independently of other people for only a very short time and rely on others for their stores and equipment. In the same way, smaller creatures live a communal life, relying on each other and on the plants that grow in their 'homes'. A meadow and the stream flowing through it, for example, provide homes for very different groups of living things. To begin with, you will probably think of plants and insects living in the meadow, and fish swimming in the stream. But if you study these two habitats more carefully, you will find that there is much more to it than that.
See how many different kinds of plants you can discover growing in the stream. Do any of them sometimes live on land?

Your project

Which creatures live sometimes on land and sometimes in the water? Do the same kinds of insects live in both places? And, most important, *why* do those particular plants, animals, insects and other creatures live in the meadow and others in the stream? Could they change places with each other? Generally, they could *not*. So we see that a meadow and a stream each has its own community — and so does a *house*.

We usually think of houses as human habitations, but in fact there is often a great deal of other animal life in a house, though the humans may not know anything about it. This book is to help you to find out.

Some of the smaller creatures that may live in a house. You could write down the names of those you know and add the others as you find out about them.

Setting about your project

There will be many ways of tackling your project but, whatever you finally decide on, your best plan is to start with a notebook in which to jot down ideas as they come to you — and of course

Your project

any questions, too — and keep it with you, because ideas may well strike you at all sorts of odd times apart from the time you set aside for definite work on your project. You might well start with a list of living things seen in your own home — and once you start you will almost certainly notice ones you hadn't seen before. If you don't know their names, make a sketch if possible or even just notes of what they look like. If you have a hand lens or one of those small 3-D viewers, you will find it very useful for watching any creatures that are new to you.

If you have friends who live in a different kind of house, perhaps in another district, you could make another list of creatures seen there. If you are fortunate enough to be able to explore a much older house, there will probably be a quite different community of living things there — and you could compare these with the ones you find in a more modern house. In its final form your project might be such a comparison, or it could be a more detailed study of a particular house. You might be able to find records of the history of your special house in your local Records Office, if there is one, and if not your local librarian could probably help you. In the records there might be interesting details about people who have lived there and even references to the living creatures they were familiar with.

You will soon realize that the kinds of life are different in different houses and at different times and you will doubtless begin to puzzle over the reasons for this. Why should *any* living creatures leave their natural home and settle in a house? ... and why *those* creatures in *that* house?

Most of the 'lower' animals — those that evolved before Man — cannot reason things out as we might. They try things out and see whether they work or not. So what must have happened is that some of them found their way into a house and soon discovered — though they would not *think* about it, of course — that they could live more easily and safely there. Then future generations would be born in the house, knowing no other kind of home. But perhaps the 'invaders' might not be welcomed by the creatures already living there, human or otherwise, and might soon be destroyed or move out of the house again.

What would the creatures be looking for when they moved into a house? Food and shelter no doubt, but what else? What conditions can you think of that would be likely to decide whether they stayed? Warmth, certainly; dryness, perhaps, though we shall see later that many kinds of creatures have to keep damp rather than dry; suitable food, of course. What else can you think of? If you were a tiny creature you would need protection from larger enemies, so that you would look for suitable hiding-places, which perhaps you would find only in one part of the house. So you see how in time the house would become the home of quite a number of families – communities of living creatures – and most of these communities would depend on each other in some way. For instance, if your mother is spring-cleaning and kills all the spiders she sees and destroys their webs, she may soon find that, instead of the spiders, she has a plague of flies and other insects that the spiders would have eaten if she had not interfered. In other words, more spiders, fewer flies.

Making your project interesting
To make a really interesting project you will need far more than just a list of names. You will find it will help you a lot if you think how to make the facts you have discovered interesting to someone else – to a friend who does not share your enthusiasms, perhaps, and to the examiner – and this is a good attitude to take up even before you start on the actual work.
Once you get really keen and interested, you are almost sure to collect more facts than you need and, in weeding out your ideas, you will find it a great help to consider the matter from somebody else's point of view. Anyone who is not familiar with the creatures you will have found out about will be interested to know what they look like, how they behave, how they can communicate with each other and perhaps with other kinds of creatures – in short, they will like to know *how they live*.

Presentation
While you are collecting the information for your project, you will be wise to give a good deal of thought to how you are going to 'present' it. Even the kindest examiner is bound to be put off by slovenly, untidy work that he (or she) finds difficult to read.

Your project

If you can type well, it will certainly be worth your while to find out whether you are allowed to use typing in your project — this looks more professional. Use double spacing rather than single — it is much easier to read. In any case, you will need to plan your margins at both sides of the paper and at the top and bottom and of course you will number your pages carefully.

You may think it worth while to use a loose-leaf system. This makes it far easier if you find you need to add something important or to rearrange any parts of your work. You will need the margins adjusted so that it is easy to read each page. At some stage, you will need to work out a definite *plan* for your work — a list of headings or divisions. Otherwise you are likely to spend too much time on some parts and not enough on others and the final result will be lop-sided.

Illustrations will need a good deal of thought. Plan them carefully, remembering that a good illustration with a brief explanation can often make your point more clearly and interestingly than whole pages of writing. They may be sketches, if you are good at that, or photographs, diagrams — whatever you most enjoy doing. A plan of the house is sure to be interesting. Do check to see that every illustration has a suitable 'caption'. You will probably refer to the illustrations in your writing, so you must number them carefully. You may think it worth while to write on only one side of each page, then you can put the illustrations in just the right places.

Finding the facts

In a project of this kind, the most valuable information will come from your own observations, but you will certainly need to read a good deal, too. Make notes as you read, if you think this will be helpful, but *don't* copy whole sentences from books directly into your project work. You will find this very dull — and so will anyone else — whereas the ideas, expressed in your own way, will fit in with the rest of your work. If you have not already a list of useful books, ask your biology teacher, the school librarian or the librarian of your public library. In this way you will avoid wasting too much time on reading books that may now be out-of-date — though it is often worth while glancing through older books for ideas or illustrations. You will find some books mentioned at the ends of the chapters of this book, but you will need to look at a good many others too.

Any museums you can visit are likely to have exhibits connected in some way with at least some parts of your subject. A single display case, if you study it carefully, will very likely suggest ideas for your project or a suitable way to present some of it.

So keep your eyes open at all times — you will find your project becoming more and more important and interesting to you, and the final result will be *your* work — including much help from various sources, but really *yours* — and that is the most important thing about a really successful project.

Houses and their human inhabitants 2

Houses and ways of life in the past

How changes in the house led to changes in the creatures living there

How people are different from the other inhabitants of houses

How life has become easier for people and other creatures

Most people in early times lived in caves or huts. As life became more settled, these gave way to houses, which gradually became larger, stronger and better planned.
Have you ever thought how different your life would have been if you had been born, say, four hundred years ago — in the time of Elizabeth I instead of the time of Elizabeth II? What differences do you think you would notice most? No radio or television so that you had to make your own amusements for the long winter evenings? Or perhaps you think you would notice more the differences in transport — no cars, or steamships or 'planes — no public transport at all — so that probably in your whole life you never went further than you could walk. There were, of course, no very large towns then, so everyone lived in the country. Very few English people traded with foreign countries, so each family grew its own food. There was no gas or electricity and so cooking took much longer and many things were eaten raw.
All these differences would make your life different. The kind of house you lived in would be different, too. What do you think houses then were usually built of? There was much more woodland, and trees like oaks provided good, strong wood for building, so in most districts houses were made of wood.
Perhaps there is an Elizabethan mansion near your home built with a strong wooden framework (so that some living creatures may actually have been carried into the house when it was built, as we shall see later). The roof would be thatched or tiled — see if

Houses and their human inhabitants

you can find out what shape it would be. Can you find out, too, what the floors were made of? It wasn't usually wood downstairs. Windows were very small and the lights were rushlights or home-made candles, so that most houses were dark inside. How was this an advantage for small creatures?

Ordinary people lived in much simpler wooden houses — the floor perhaps made of earth trampled down — and few of these houses are left today. So the earliest *real* house you are likely to find is an Elizabethan mansion.

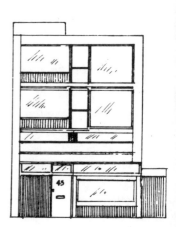

A very modern home and two earlier ones. Which home do you think had the most living creatures in it and which one had the most different kinds?

Changes in the house and its inhabitants

Any surviving Elizabethan mansion has almost certainly been 'modernized' at various times since it was built. When it was about a hundred years old, the stone floors were very likely to have been replaced by wooden ones, and these did not need the rushes which had covered the earlier stone ones. This meant that living creatures were no longer brought into the house among the rushes and a useful hiding place was lost. At a later date, one very large room would be replaced by several smaller ones. Each of these rooms by this time had its own coal grate where some small creatures could find shelter; earlier, the open hearth in the middle of the floor had not been so useful to them. When, nearer our own time, grates were replaced by gas or electric fires, these hiding-places vanished. If central heating has been installed more recently still, this will have dried out the house and destroyed the damp corners where spiders and some other creatures love to lurk.

Houses and their human inhabitants

If the roof was originally thatched, there would have been lots of insects and spiders and perhaps mice or even rats, as well as birds, living in the thatch, especially if it wasn't repaired very often. All these creatures would be driven out and would have to find somewhere else to live when the thatch was taken off and tiles put on.

Until about a hundred years ago, there was no running water in the house. When the owner decided to be up-to-date in this matter, tanks and pipes would be brought in. Small forms of animal life were undoubtedly living in the water and so would be brought into the house, though they might not survive for long. One of the places in a house where you may find a number of different kinds of living creatures is the bathroom. I expect you can work out why these are different from those you find in other rooms. And of course the bathroom isn't used for much of the day, so the creatures who live there stand a good chance of not being discovered or disturbed too often.

When rubbish was not collected and disposed of regularly there were far more germs than there are today — and germs of course are tiny living creatures. Think, too, of the difference it made when food could be stored in refrigerators and metal cupboards instead of wooden ones. Which small animals do you think suffered by losing their food supply in this way? Can you think too, which insects could no longer find food to lay their eggs on? All these changed conditions in the house were caused by the people who lived there. Naturally they were the most important form of life in the house, because it had been planned and built to meet their needs. The other creatures just took advantage of the food and shelter they could find there.

If your special house is an old one, you can think of the various owners living there one after another and imagine the differences they must have caused with their various new ideas. Some of the things they did are very likely still making a difference to the

In which of these fireplaces could some small creatures find shelter? Do you know what kind of fuel was burnt in each fireplace?

  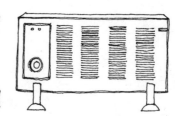

people and other creatures in the house now. Can you think of some?

I think it will help you to sort out your ideas for your project if you can decide now which 'special' house you are going to deal with in it. (Of course there is no reason why you should not mention other houses if you wish.) Then you can draw a good, clear *plan* of this house or flat, and this will form a sort of reference map for your work. As we think about the various kinds of living creatures you can think where they might have lived in *your* house — or you may be able to discover where they *did* live — or do now. If you put a coloured dot — or some other symbol — on your plan for each of the different kinds of creatures that lived there, you will begin to notice which ones lived with each other and be able to work out some of the reasons for this.

Even if you are not specially studying a modern house, it will be a good idea for you to take a paper and pencil and go slowly round the various rooms in your home making a list of the creatures you find in different kinds of places. Good places to look will probably be behind pictures, under rugs, in the corners of cupboards and among the folds of curtains and other hangings. You yourself will probably think of other hiding-places. Don't leave out any of the creatures you find because you do not know its name — there are plenty of books with clear drawings, from which you will be able to name them, especially if you make a sketch to show the shape or anything special you notice — very long legs or antennae, for example, or one that looks like an insect but has not six legs, as all insects have.

People compared with other inhabitants of houses

To return to the people in the house, it will be a good idea, before we go any further, to think how humans are different from all the other creatures you are likely to find. Let us make a list of all the ways we can think of.

1 One way we have already mentioned — people can *think* about things; most animals cannot. They live by *instinct*. This means that from the time they are born they will act in a certain way if they are faced with a certain problem. If young birds are hungry, they will shriek for their parents to feed them and they may keep on, even if it is dangerous to do so because their

Houses and their human inhabitants

enemies may hear their calls. In the same way a frightened animal will 'run for its life'. By acting in these ways, the creature probably saves its own life, but, as we shall see later, there are times when it would get on better if it could reason things out as people can.

2 Another way in which creatures differ from us is that, since they do not think, they cannot remember the past or make plans for the future, as we do. So their life is lived only in the present.

3 I expect you have realized that other creatures do not know *why* they do any of the things they do. In fact, all their actions are controlled by two things — the wish to stay alive and the urge to have young and so keep their race alive. Of course, these are important to people, too, but this is by no means all that interests them.

4 We shall see presently when we are thinking about various other creatures that their senses are different from ours. We ourselves live in a world of sights and sounds, but for other creatures smell or touch may be more important, and this means it is very hard for us even to imagine their way of life.

See if you can think of other ways in which we are different from other creatures. If you write down in your notebook everything you do in one day from the time you get up until you go to bed, I think you will be interested to see how many of these actions one of the flies or spiders or other creatures in your home cannot do — or does not need to do. It will be interesting, too, to notice how many of your actions are necessary to keep you alive and how many are not.

How life has become easier

In very primitive times, *everybody* — man, woman and child — spent all their time getting food, and even then they often went hungry. Later on, there were better ways of farming and so one man could produce enough food for himself and two or three other people. Nowadays, by using machines and electricity, one man can provide food for about twenty people besides himself. This means, of course, that the other people can work in shops, factories, offices, schools, buses and other places and so make life easier and more comfortable for them all. So today people can provide food and shelter for large numbers of other living creatures in their homes, and not even notice it. In later chapters we shall find out something about these other living creatures that people have in their homes — or had in the past — either deliberately or unintentionally.

A helpful book
The Living House by George Ordish (Rupert Hart-Davis, 1960), the history of Barton's End, a four hundred-year-old farmhouse in Kent and of the various living creatures that are known to have lived there. It has clear black-and-white drawings of many of the creatures.

Pets and other mammals 3

Pets today and in the past — dogs, cats, birds and crickets
Other mammals — rats, mice and bats
Why these creatures like to live in a house
Adding them to the plan of your house

The larger animals that people deliberately have in their houses nowadays are of course their pets. So many more people can now afford the time and money for pets, that there are far more of them and many more different kinds than ever before. You could make an interesting section of your project by making a Pets' Corner — pictures or photographs or sketches of as many different kinds of pets as you can think of. You could add interesting notes about the best way to keep these pets in a house or garden. The most popular pets today are dogs and cats; these are also the ones that first came to live in houses. In the past, a good many famous people had their special pets. Dr Samuel Johnson, who published the first English dictionary in 1755, used to go out and buy oysters for his cat, Hodge. Lady Munnings, the wife of the President of the Royal Academy a few years ago, had a Pekinese called Black Knight, who was the only dog ever to be made a Freeman of the City of London. Her biography of him, called *The Diary of a Freeman*, made enough money to buy an ambulance for the P.D.S.A.
The very first dogs that came to live in or near houses long ago were those that men had trained to hunt for them. Earlier still, wild dogs had followed the huntsmen around, hoping to pick up scraps left over from the animals men had killed for food. Gradually some of these dogs came to trust people enough to live with them and work for them in various ways. You could illustrate ways in which dogs work for people now — or did in the past. See if you can find out how one kind of dog actually helped to cook the joint. Perhaps, too, you have seen sheep-dog trials,

or seen them on television. What else can you think of?

If you have ever had a dog and a cat as pets, you will know that they look on humans quite differently. A dog seems to look on his master as a god and takes it for granted that what his master says and does is right. Wild dogs lived and hunted in packs, you see, and the dog's master has taken the place of the pack-leader. A cat is quite different. He depends almost entirely on his master for food and shelter and sometimes allows himself to be petted, but he is never so completely dependent on people as a dog is. Do you see why this is so? Wild cats live alone, hunting at night. In earlier days, men valued them because they caught rats and mice and so prevented their houses from being overrun with them, but cats were never so useful to people as dogs were.

If you have watched cats at all, you will know how they love to be *comfortable*, and this is what attracted them to men in the

first place — especially the fact that men could make fires.

You know how a cat loves to lie in front of the fire. What was it that attracted people to cats, do you think? Their soft, warm bodies or their bright eyes, or what? Often, as in ancient Egypt, cats were looked upon as sacred animals and played an important part in religious ceremonies. So early man seems to have had a kind of respect for cats that would make him almost feel that it was an honour to have a cat in his house.

I wonder if you have seen your cat behave like the one in this poem by Harold Monro:

Pets and other mammals

Milk for the cat

When the tea is brought at five o'clock,
And all the neat curtains are drawn with care,
The little black cat with bright green eyes
Is suddenly purring there.

At first she pretends, having nothing to do,
She has come in merely to blink by the grate;
But, though tea may be late or the milk may be sour,
She is never late.

And presently her agate eyes
Take a soft large milky haze,
And her independent, casual glance
Becomes a stiff, hard gaze.

Then she stamps her claws or lifts her ears,
Or twists her tail or begins to stir,
Till suddenly all her lithe body becomes
One breathing, trembling purr.

The children eat and wriggle and laugh,
The two old ladies stroke their silk;
But the cat is grown small and thin with desire,
Transformed to a creeping lust for milk.

The white saucer like some full moon descends
At last from the clouds of the table above;
She sighs and dreams and thrills and glows,
Transfigured with love.

She nestles over the shining rim,
Buries her chin in the creamy sea;
Her tail hangs loose; each drowsy paw
Is doubled under each bending knee.

A long, dim ecstasy holds her life;
Her world is an infinite shapeless white,
Till her tongue has curled the last holy drop,
Then she sinks back into the night,

Draws and dips her body to heap
Her sleepy nerves in the great arm-chair,
Lies defeated and buried deep
Three or four hours unconscious there.

Perhaps you would like to quote this poem, or another you have found, in your project; it would be very effective if you could decorate it with little sketches. Or perhaps you think it would be a good idea to put in sketches or photographs of your own cat. Ones that show the cat doing something will be better than those where he (or she) is still. You may like to do the same for your dog or other pet.

In the past, people often kept birds in cages in their homes because they liked to hear them sing. Often these were wild birds who soon pined and died; today, of course, no one is allowed by law to cage a wild bird. Some people in the past even kept crickets as pets, because they liked to listen to their song. Crickets are chirping insects rather like grasshoppers. Have you ever heard crickets chirping? If so, you can decide what you think about the idea of having them as pets.

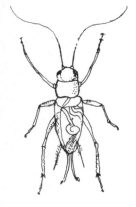

A house cricket.

If you have crickets in your house and are able to watch them, you will know that their 'song' is not singing at all, in the way that people sing. See if you can discover which part of their body they use for this. You will find that it is the forewings that are rubbed together — something like rubbing two files on each other! And do you know how they 'hear' each other? — for that is the real purpose of the chirping of course. They do not have ears like ours, but there is a patch below the 'knees' on their fore-legs which can catch the sounds from other crickets.

Other mammals

Most of the pets that people keep are mammals — you can check this word in your dictionary. Some other mammals that live in houses, but not usually as pets, are also described as rodents. What does this mean? Rats and mice are the chief rodents to be found in houses. There are three ways in which these rodents can do harm to people. One of these ways is connected with human food and one with the house itself. Also, as we shall see later on, rats and mice carry disease.

Some recent research about mice has brought out some very interesting new ideas about these creatures. Did you know, for example, that the sole of the mouse's foot has glands which send out an oily substance? What do you think is the purpose of this? You have very likely heard about birds that have their own

Pets and other mammals

'territory', a piece of land where only one pair of birds is allowed to come during the breeding season; others of the same kind of bird are 'warned off' and even driven off. It now seems that where there are a lot of mice together, each male has his own territory in the same sort of way. Male mice can and do fight very fiercely. I expect you have heard, too, about the 'peck order' among hens. The cock, of course, rules the roost, but among the hens some are fiercer and stronger than others and peck those that are weaker. So an average hen can expect to be able to peck half the others and to be pecked by the other half. There is the same kind of arrangement among mice. What does it mean when we say a man is hen-pecked?

A field mouse feeding.

Pets and other mammals

A long-eared bat flying.

Did you realize that bats are flying mammals? They, too, sometimes shelter in houses; they did so even more in the past when the roofs of houses were not so strongly built as now; they like to rest in an attic during the day and fly out in search of food at night. If you have ever watched bats flying around, especially in rather a small space, you will have noticed how skilful they are at avoiding any obstacles in their path. We now know that bats have a kind of built-in radar to guide them. They can also 'talk' to each other by means of sounds that are far too high-pitched for us to hear.

Why these mammals like to live in houses

All the mammals that share man's home with him need the same kinds of food as he does, so it is easy to see what has attracted them to houses. Of course, people can now store their food in much safer places than in the past, so that it is kept for themselves and the creatures they wish to feed; pests like rats and mice do not find it nearly so easy to get at. *Water* is even more important than food for all these mammals, so that this attracts them to houses more than food does.

Pets and other mammals

Marking mammals on the plan of your house

I wonder how you will decide to mark in these mammals on your plan? This needs a little thought. Don't forget, for example, that rats and mice have not only holes but also runs, paths that they follow regularly. If there are still any of these creatures in 'your' house, as there well may be if it is an old one — they seem to come back time after time, following their old runs — you may be able to find just where the run leads to.

Cats and dogs, too, like to have their own special places, marked by their own particular scent, which is more important to them than things they can see. If you have more than one cat or dog, it will be interesting to notice whether their special places are close together or not.

Since mammals are fairly large, a tiny sketch of each one would look very well on your plan. Then you could use coloured dots or other signs for the smaller creatures we shall be thinking about later. If you have already marked on your plan the places where there is food or water, you will find the whole thing is beginning to make sense.

Books

1 *An Animal Lover's Scrapbook* by Patrick Ney (Max Parrish, 1963).

   The author has collected newspaper cuttings for a long time.

2 *Mice All Over* by Peter Crowcroft (Foulis, 1966).

   This is an account of a piece of research on the house-mouse. The research was carried out by a scientist in order to find out what damage these mice do to stores of grain. It gives a good idea of how scientists set about their research; if you are at all interested in mice, you will find it very good reading.

4   Birds

How some birds came to live near farmhouses and even in busy cities

House-martins and house-sparrows are often enemies

Other birds that live close to houses now

How birds help people

In the last chapter we spoke about song-birds that people kept in their houses, but, because they can fly, wild birds don't *have* to live anywhere near houses. Yet, as time goes on, more and more of them seem to be coming to live quite close to houses and even occasionally venturing inside, of their own free will. Let us see how this has come to be.

Birds that live on or near farmhouses all the year round

The lives of many birds were altered when the forests which covered much of England in early times were gradually cleared

Birds feeding on the farmer's crops. Do you know their names?

Birds

A cock robin feeding his mate.

away. This was sometimes because the timber was needed. You could make a list of things it was used for. I expect you have heard how it was used at one time for smelting iron — what else have you thought of? At other times the forest was cut down in order to make space for farming.

When forests were cleared, birds, like jays, that can only live among trees, disappeared from the district. Other birds, like green woodpeckers, however, might have gone away for a time and then found that, after all, there were enough trees left for their needs. There might also have been a good supply of food, such as ants, in the fields that had taken the place of the woods. The farmer's crops also provided food for many birds. In fact, in time, some birds, such as pigeons and magpies, did so much damage to crops that farmers had to shoot some of them to keep the numbers down. On the whole, however, the damage that birds may do to crops is much less than would be caused by the insects that the birds eat.

So, although the cutting down of the forests drove some birds away, others soon took their place on farmland and very soon some kinds came to live around the farmhouse.

Birds

Here is a list of some of the birds that are often seen on farmland. See if you can find out whereabouts on the farm you are most likely to see them. Do they feed on any particular plants? Do they nest and roost in trees? Do they live in pairs or in flocks?

Rooks, jackdaws and magpies   Starlings and linnets
Goldfinches and chaffinches   Pied wagtails   Blackbirds and thrushes   Robins and wrens   Dunnocks — they used to be called hedge-sparrows, but they are not really sparrows at all. You might add to the list other birds you have seen yourself on farmland, especially near a farmhouse.

If you do not know these birds, there are plenty of interesting books that will show you what they look like and tell you a good deal about the way they live, but of course the best way to get to know birds really well is to find out for yourself by watching them. If you have ever tried to stalk a bird, you will know that their eyesight is so keen that they immediately spot any movement that might mean danger to them. If you move very slowly and smoothly, however, you can sometimes get quite close up to them before they fly off.

You could make quite a good start in your own garden, very likely, especially if you put out the kind of food that garden birds like. Bread, seed and nuts are all likely to attract them. It is safer to put the food on a bird-table rather than on the ground. You could ask your woodwork master about making one. If you want to attract birds like blackbirds and thrushes, that will not fly onto a bird-table, you can put the food on a tray on the ground and take it in at night, so that you do not also attract rats. If you have no garden, you could still attract birds to your window-sill if you put up hooks around the window and its sill and hang food on them.

You are sure to see plenty of house-sparrows, even if you do nothing special to attract them, and you will perhaps be surprised to discover how very distinctive and beautiful their markings are. Look for two that have exactly the same markings. It is quite difficult to find this.

Migrant birds that live near farmhouses
Migrants are birds that live in one country during part of the year but spend the rest of the year in other lands — usually warmer ones.

Birds

Among the migrants that visit England in the summer are swallows and house-martins. These birds nearly always live around farm buildings. Can you think of something else besides the cold weather that makes it impossible for these birds to stay here in the winter? They nest on or in the buildings and feed on the insects they can catch. Swallows seem to prefer to build their nests in barns, but martins choose houses. Both make mud nests, so you can see why they look for a place near a river where there is plenty of mud and insects. Do you see why they choose a place fairly near the coast if possible?

If you can watch these birds, you will notice how carefully they choose the mud for their nests. How do you think you would choose your mud, if you were one of these birds? If you are learning pottery, that will help you to think what you need to look for. The nest must be strong, of course, but also it must be a good shape, so that it will not crack in the sun. You may be able to see what it is that the birds work in with the mud to make the nest stronger. You will notice that it is no good trying to watch them in the afternoons. Why do you think it would be a bad idea to put on too much damp clay at a time?

When it is finished, a house-martin's nest looks very much like the shell of a coconut with the top cut off, for it is the same size,

A pair of house-martins on their nest.

shape and colour. You should look for these nests at the top of the wall of a house right under the overhanging eaves of the roof. What is the advantage of this position, do you think?

House-martins usually arrive in England from South Africa about the middle of April. It has now been proved by 'ringing' some of the birds (i.e. placing a very light numbered metal ring round one of the legs, so that the bird can be recognized again), that the same martins return year after year to the same place. They immediately set about repairing their old nest or building a new one if sparrows have taken over the old one and the martins can't drive them away. Incidentally, it may sometimes be doing the martins a kindness to destroy their old nests. After a time they usually contain the eggs of mites, bugs and lice, and the insects that come from the eggs suck the blood of the martins, especially of the young ones.

Martins do not 'keep themselves to themselves' in the breeding season, as a great many birds do. Several martins may build their nests quite close together on the same house. They go into each other's nests and flocks of them may be seen flying around 'hawking' for insects. If they can find a nice sunny spot, whole flocks will settle there to enjoy the warmth. If there are martins near your home, try to watch them doing all these things, and also building their nests. What else can you see them doing? I am sure you will find it worthwhile to keep a diary of their activities. You will probably be surprised to find how much you notice — and how much you have missed before.

Birds that may live in towns

House-sparrows have now become so accustomed to people and their way of life, even in crowded cities, that many of them depend entirely on people for their food and do not go back to live in the country if food becomes scarce, but simply look for another house where they can find it. Within the last few years, jays and magpies have been seen much more often right in the centre of London, so it looks as if they, too, are learning to depend almost entirely on people.

Pigeons and starlings, too, often live in big cities, though there are fewer pigeons since there have been cars and vans instead of horse-drawn vehicles. Do you see why? Pigeons mainly

Birds

eat corn. Think of what horses used to eat. They carried their food in a nosebag. Do you know what these were like? Your grandparents could tell you about them.

Birds on the roof of a house. What are they doing?

Rivalry between house-martins and house-sparrows

Because of their swift flight, martins can escape from birds of prey which might otherwise kill them, so that their only real enemies near houses are cats — and sparrows. As we have seen, sparrows may steal their nests — they seem to be too lazy to build their own nests — and they may even throw out the martins' eggs. It is a curious fact that nowadays the two birds that most often live close to houses, the house-sparrow and the house-martin, are very often enemies. Birds often become enemies because they want the same food as each other, but this is not the case here. The natural food of sparrows is grain and seed, while martins feed on the insects they catch as they fly along. Because of this difference of food, you will see that sparrows have short, stout beaks (for cracking seeds), but martins have much thinner and longer ones. You could make sketches to compare (a) their beaks, (b) their wings and (c) their legs, and explain how each bird is fitted for the kind of life it leads.

Sparrows are bold and soon cease to be afraid of people. When a new house is being built, nowadays, the sparrows often appear on the first day and eat any crumbs the workmen drop at lunch time. Have you seen this happen? Originally they ate grain and the seeds of weeds and some caterpillars and butterflies. But as time has passed, they have become more and more dependent on people, until now they prefer to hunt among household rubbish for their food, or perhaps take any food the dog has left in his bowl. They very quickly spot food put out on a bird-table. It is interesting to notice, however, that adult sparrows still eat some insects and they feed their young entirely on insects.

What else can you think of that both sparrows and martins might want? As we have seen, sparrows often steal martins' nests, so it seems as if it may be *space* that they are both wanting. Sometimes perhaps there isn't enough for all the sparrows and all the martins who would like to live on a particular house.

While the martins are at the other side of the world in winter, the people in the house may feed the sparrows, especially if it is very cold weather. This gives the sparrows a better chance, while at the same time there may be less food for the martins because modern farmyards are much cleaner than they used to be and everything possible is done to drive away flies and other insects from the farmyard and the milking sheds. At the same time farmers are now spraying their crops to kill insects that they regard as pests. So again there are fewer insects for the martins to eat.

Other birds that live near houses now

Apart from house-sparrows and house-martins, which actually nest on the house itself, starlings may nest on creeper growing on the walls of the house, while jackdaws sometimes build their nests in the chimneys and so cause a lot of bother when fires are lit in winter. Many other birds may be seen roosting on the

Birds

roofs and chimneys of houses. I expect you have heard about the huge flocks of starlings that fly to Trafalgar Square every evening to roost on the buildings there. Perhaps you have seen them there or on their way there. In fact, there are now so many of these starlings that sometimes people are obliged to put grease on the window ledges to prevent the starlings from alighting.

A great many people, too, used to feed the pigeons on Waterloo station in London, just as they do in Trafalgar Square or in front of St Paul's Cathedral. But at Waterloo the pigeons have now become such a nuisance that people have had to be forbidden to feed them.

When houses were thatched, tits were often seen hunting among the thatch for insects. They are such tiny birds that they can squeeze into a hole only about an inch across, so it is often easy for them to find a place for their nests on a house, if some part of it is wooden. You will find that tits will not use a nestbox with a larger hole. Do you see why?

Since people have had their milk delivered in bottles, the tits have discovered that they can peck through the foil tops and

A great tit on a milk bottle.

drink the cream, which they love. They will almost certainly come to your garden if you put out monkey nuts and pieces of fat. So you see they too are coming to depend more on people. In fact even shy birds like these have sometimes been tempted indoors and lived there quite happily, coming and going as they please. Some interesting tests have been carried out on tits to try to find out whether they can see colours. Milk bottles have been put out with different coloured tops and watch has been kept to see which ones the tits would peck. It is almost certain that they *can* distinguish different colours.

People who feed birds in their garden find from time to time that they are attracting a new kind of bird. Just lately this has happened with a lovely bird, something like a slender pigeon, called a collared dove. In a very small garden I know on a busy road quite near London there are always crowds of sparrows, starlings, greenfinches, chaffinches and tits, and some robins, thrushes, blackbirds, pigeons and now collared doves.

Linocuts of some of the birds mentioned in this chapter, particularly those that live near to houses, could look very well in your project. If you can find a bird's feather, and look at it carefully, that makes a good linocut, too.

How birds are useful to people

Ralph Hodgson in this poem points out how much *people* may depend on *birds*:

Stupidity Street
I saw with open eyes
   Singing birds sweet
Sold in the shops
   For the people to eat,
Sold in the shops of
   Stupidity Street.

I saw in vision
   The worm in the wheat,
And in the shops nothing
   For people to eat;
Nothing for sale in
   Stupidity Street.

Birds

(In fact, of course, horrible as it may seem to us, lark pie was regarded as a very great delicacy in earlier times. There is even a recipe for this in Mrs Beeton's Cookery Book published in 1870.)

To sum up, we have seen that more and more birds are coming to live near houses, or even in them, because they find good places to build their nests and can get food there more easily than elsewhere. It is important, too, to remember that birds are a very great help to people, especially farmers. This is because of all the insects the birds eat, which would otherwise do a great deal of damage to crops and fruit trees and so give people less to eat.

Interesting books about birds
1 *Living with Birds* by Len Howard (Collins, 1956).
   Stories of the birds, mainly tits, that the author made friends with in her cottage in Sussex. Many of them she got to know individually and even taught one to count to five!
2 *The Bird Table Book* by Tony Soper (David & Charles, 1965).
   About feeding, encouraging and enjoying wild birds in the garden and on the window sill.
3 *Sold for a Farthing* by Clare Kipps (Arthur Barker, 1953).
   The story of a sparrow called Clarence, who lived with the author in her home because he had been injured very early in his life.
4 *Bird* by Lois and Louis Darling (Methuen, 1963).
   You may find some of the text rather difficult, but do try to get a copy and look at the illustrations, which are delightful.
5 *The Life of the Robin* by David Lack (Penguin Books, 1953).
   As well as telling about robins, this book describes a number of experiments carried out with these birds, especially to find out how they will behave in various circumstances.

5 Insects

An insect's way of life — its instincts and its senses

Insects in houses: (a) those that bore into wood
(b) those that live in dry places

How insects and other creatures affect each other

We can *describe* the life of an insect, but it is almost impossible for us really to know what it would be like to *be* one. If you make a list of ways in which insects are different from people, this will help you to imagine an insect's way of life more fully and exactly. You could do this as you study this chapter and add ideas of your own — differences in size and in ways of getting about, for instance.

Then, too, in the course of its life-time, an insect lives through several different stages. I expect you have seen this in the case of a butterfly; the egg develops into a caterpillar (or larva), which later becomes a pupa (or cocoon) and in the end the adult butterfly. The caterpillar crawls around, although the adult butterfly will be able to fly. It would be interesting to make coloured drawings of a particular kind of butterfly — the special food-plant with eggs on it, the caterpillar crawling and eating, the pupa resting and the butterfly in flight. Best of all would be to do this from the real insect, but films, books and museum displays will help you.

Many insects have to start life in the water, although later on they live on land or in the air. Can you think of some that do this? Coloured drawings would be good here, too, or you might make a small panel in needlework — cross-stitch on canvas, perhaps, — your needlework teacher would set you off. If possible, find a needlework picture done in the past, two or three hundred years ago. Such pictures nearly always show insects and caterpillars and other creatures and would give you some ideas. Look in a museum or a large country house.

Some artists seem to have enjoyed painting insects, too. If you

Insects

look in an Art Gallery or in a book about painting, you may find some paintings by the artist Crivelli. Whatever the picture is about, it is almost certain to include some insects.

A piece of embroidery worked three hundred years ago. How many living creatures are there?

The instincts of insects

In spite of the different forms it takes and the different places it may live in at various times in its life, an insect's life is really quite a simple one ruled by instinct. This means, as we have seen, that if it is faced with a particular problem the insect will respond *automatically* in a certain way. It does not work it out; it just knows what to do. For example, perhaps you have seen or heard about the hunting wasp. This insect digs a burrow, finds a suitable caterpillar, paralyses it with its sting, drags it to the burrow and lays its eggs on it. In this way, the wasp-grubs will have a suitable meal waiting for them when they hatch out. Yet the wasp does all this instinctively, without knowing why it does any of it. Insects, like most small creatures and even birds, live in the present with no memory of the past and no idea of the future.

What happens, do you suppose, when an insect is faced with a new problem, so that it does not know what to do? It cannot reason it out, as we might. Can it learn in any way to solve the problem, do you think? The answer seems to be that *some* insects can *sometimes* learn to do simple actions.

There have been a number of experiments to find out what happens to a bee or wasp if it is taken some distance away from its nest. Will it be able to find its way home? After watching these insects very carefully for a long time, scientists have found that they sometimes can. This is because they rely on certain landmarks — trees, bushes, stones and so forth — both near their nests and between there and their feeding grounds. If they have been taken away from the nest, they will fly all round until they find some of these landmarks and then find their way home from there. By moving the landmarks, however, it is very easy to mislead the insects, so that it is clear that they do not understand what they are doing. They are still relying on instinct.

Even insects like ants and honey-bees, which live a very complicated social life and certainly communicate with each other, still depend entirely on instinct. You could show this idea very strikingly if you could make a pen-and-ink cartoon to contrast the supposedly *reasonable* behaviour of a person or group of people with the *instinctive* behaviour of an insect — a hunting wasp, a bee, an ant or any other you are interested in. Finding a good title for your cartoon should be fun.

An insect's senses

Which senses do you think are most important to an insect? Sight, hearing, taste, smell or touch?

If you can examine the eye of a house-fly under a microscope, or look at an enlarged picture of it, you will realize that, wonderful as our own eyes are, those of an insect are even more marvellous. You could make a list of reasons why an insect needs

Insects

such very keen eyesight, and perhaps a list of insects that you think can see things you can't, because their eyes are more sensitive than yours.

How do you think you could test how well an insect can hear? First, of course, you will need to discover its ears. Remember that the 'flap' of the ear which is so noticeable in people and other mammals, is only used to *collect* the sound waves. The real hearing takes place *inside* the ear. What you need to discover is whether your insect shows at all that it feels vibrations caused by sound-waves. You may even be able to show that it can 'hear' sounds that we cannot because they are too high or too low. If you have done any experiments in your science lessons, you will know that you almost always need a 'control', to make sure that when the insect moves, for example, it does so because of something it hears and not because of something it is aware of by sight or another sense.

You could carry out tests about the other senses, too. Your biology teacher would tell you which insects would be best to test and whether the kind of test you have thought of is likely to give you a good result.

When you have found out something about its senses, you will be a bit nearer knowing what it would be like to be an insect.

Why insects quickly get used to living in a house

In the course of evolution, it sometimes happens that one creature is in some way different from all the others of its kind. If this change makes the creature more successful – in getting food, or in finding a mate or a good place to live, perhaps, – then this new characteristic will be passed on to the offspring. Insects multiply very rapidly; a greenfly born on Sunday may be a grandmother on Tuesday. Then, too, they live through their life-cycle in a very short space of time. So any change is very soon passed on and after a very short time *all* the insects of that kind will show it. This is why many insects are able to make a success of living in a new place, such as a house.

Two kinds of place in a house that insects find specially attractive are: 1 In the wood of the house itself, or of furniture, and 2 Among dry litter.

Wood-boring insects

Wooden houses in the past were usually built slowly by hand, so that it often happened that the trees were felled and then just left lying about until there was time to drag them away and saw them up ready for use.

In the wood might be some *Ambrosia beetles*. Perhaps you have heard them called 'pinhole borer beetles'. What happens is that

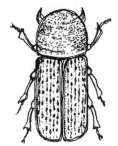
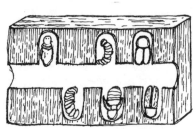

The Ambrosia beetle bores deeply into wood. You will notice that its brood-channel has separate places for grubs, pupae and beetles.

a fertilized female beetle finds a suitable tree and starts to tunnel into it and then lays her eggs in the tunnel. The next step is the most interesting. She places around the eggs a certain kind of fungus. (How does she know that this is a good idea, and which kind of fungus is needed? Again by instinct.) When the eggs hatch, the larvae feed on the fungus, which continues to grow so long as the timber is fairly damp. The fungus partly dissolves the wood; otherwise the larvae could not digest it. Finally the larvae travel back along the tunnel made by their parents and so reach the outside of the tree, ready for their life as adult beetles. When the wood is used for building a house, there may be eggs or larvae already in it.

Another creature that sometimes attacks either the wood in the house or the furniture is the *furniture beetle*, which is often spoken of as a woodworm. In this case the eggs are laid in cracks in suitable wood. The young grubs, as soon as they hatch out, immediately begin to bore into the wood. These grubs are so tiny that our eyes cannot see the holes they make when they go into the wood. What we *do* see, when wood is 'worm-eaten', are the rather larger holes made by the adult beetles when they come to the surface after spending some years inside the wood. Naturally, during this long period, they have been burrowing into the wood and feeding on it, so that it often becomes rotten and useless.

A furniture beetle.

Insects

A third wood-boring insect is the *Death Watch Beetle*, which often does great damage in old churches, but may sometimes be found in the wood of houses, too. These beetles only attack dead wood. If you go to a cathedral or large church where this beetle is causing a lot of damage, you will probably find that there is an appeal for funds to stamp it out and replace the rotten wood. If so, there are often enlarged photographs of the beetle and how it damages the wood. These would be very good to copy.

The adult beetles are ready to crawl out of the wood in May or June. Just before they do so, they make a peculiar tapping sound, very like the ticking of a watch. To do this, they strike their heads down on the wood seven or eight times in a second and then stop. Soon the taps are repeated, probably from a spot nearby. This is another beetle answering the first. I expect you realize that these taps are the signal by which the male and female beetles find each other.

It is easy to see what an eerie sound this tapping may be if it is heard in the stillness late at night, so that it is no wonder that quite a lot of people believe that if they hear the sound it means that someone in the house is going to die. So you will realize how these beetles got their name.

Insects that live in dry litter

If you take a walk anywhere you will come across plenty of *wet* litter, but very little *dry* litter. It would be interesting to check this by noting where you find each kind and how much of it.

Early man liked dry places — caves and, later, huts. He liked to keep them as warm as possible, too. At first, of course, they were very dirty, so that there was a good deal of litter in them, but it was *dry* litter. What was in this litter, do you think?

After a time, people started to wear clothes; furs and skins and later on woven woollen cloth. Lots of small creatures lived at first among dry material like moulted feathers and hair, cast snakes' skins and even dry dung. Later they lived among the clothes that people had stored away, perhaps in wooden chests. They had to find somewhere *dry* if they were to survive and flourish.

Chief among the insects that flocked to man's clothing, of course, was the *clothes-moth*. This moth will eat, and lay its eggs among, clothes made of wool, leather and silk, and these were the chief materials used for clothes until fairly modern times.

We hear of Elizabethan housewives trying to get rid of the moths by brushing the clothes, airing them in the sun, seeing that they were clean before they were put away and placing sprigs of lavender and laurel among them. Perhaps your mother does some of these things.

Since clothes-moths love dry, warm places, there were more and more of them as time went by and ways of heating houses improved. But now things are changing. With central heating, the air is becoming *too* dry for them and also carpets and clothes can now be made moth-proof and clothes are often made from man-made fibres, such as nylon, that moths do not like. You can probably think of other ways in which moths are got rid of now, but in an older house you could very likely find the moth, its caterpillar and its eggs and make sketches of them.

Two other interesting little insects that may be found in dry spots in houses are *silverfish* and *firebrats*. Both of these are insects without wings that have survived from very early times almost unchanged.

The adult *silverfish* is about half an inch long, shaped something like a carrot, and covered with silvery scales. A long time ago, these creatures had hairs instead of scales. The scales must be useful to them in some way that hairs were not, but at present no one seems to know how. Silverfish like to be in the dark, so that if you find any in your house they will most likely be scurrying for shelter and you will have to be quick to spot them. If you can get near enough, you will see that they have a pair of long antennae (feelers), three bristles at the rear and a pair of compound eyes.

These small creatures have discovered a good food-supply among books. They scrape away the paper and eat the starchy paste underneath and also the glue and size used in the bindings. They do not have such an easy time now because modern book-binding glues and pastes are made from imitation resins, which silverfish do not like, and also there are insecticides which will kill them. Although silverfish like to live among dry litter, they often find it difficult to get enough water. So you could look for them also in damp places in your house — under sinks or behind wallpaper.

Insects

A silverfish.

Firebrats are rather like silverfish, but they seem to be able to live in any warm place, such as under stoves or, in earlier times, behind firebacks. This love of warmth, of course, has given them their name. They abounded in the kitchens of earlier houses and would come out at night to feed on the scraps of food left about on shelves or on the floor. Modern kitchens, with their gas or electric cookers, do not provide homes for these insects and it is unusual to find scraps of food lying about in kitchens nowadays.

Various insects were common at one time in stores of food that were kept in houses or close to them. These included *flour-beetles*, the *Indian meal moth, grain-* and *rice-weevils* and *fruit-moths*. With modern storage methods, these insects have almost entirely disappeared, but you might enjoy hunting for pictures of them and information about them, perhaps in old books of household hints.

A firebrat.

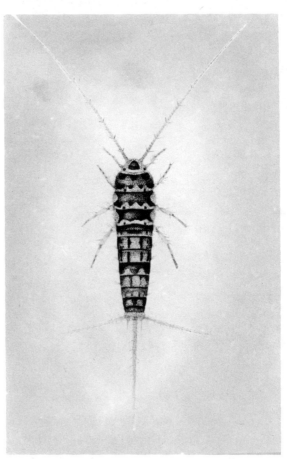

Among the books in a house there may also be booklice, bookworms and sometimes clothes-moths. *Booklice* are very much like the lice found on birds, so there is little doubt that they lived at first in the thatch, then in birds' nests and then inside the house. Some may also have been carried in in the books themselves. They like to feed on moulds and other fungi, so that they are usually found among damp and mouldy books. If there are males, they make tapping sounds something like the Death Watch beetles. These are the mating calls.

Bookworms are sometimes also called bread beetles. Like silverfish, they go for the starchy paste in books — and of course

Insects

bread is starchy, too. The eggs are laid on the outsides of old books; the larvae burrow right down into the pages, then turn into pupae and later into adult bookworms. These burrow their way out again, fly around and find a mate. Then the whole cycle starts again. We sometimes speak of a person as a 'bookworm'. How much is he, or she, like the real bookworm?

Insects compared with other creatures

There are more insects in the world than any other kind of living creature. In Britain alone there are more than twenty thousand *different kinds*. Most of these insects produce thousands of eggs during their lifetime, but only a few of these survive long enough to complete their life-cycle. This seems very wasteful, doesn't it? It *is* wasteful from the point of view of that particular kind of insect, but from the point of view of life as a whole there is very little waste. You probably understand why this is so. For example, insects feed on each other, spiders feed on insects, birds feed on both, cats feed on birds and so on. A number of living creatures feeding on each other like this is spoken of as a 'food chain', and every living creature has its place in one food chain or another.

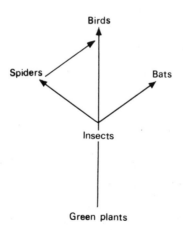

Part of a food chain. What do you think would feed on the birds? And on the bats? Or are they at the end of the chain?

Most important of all, every kind of creature, including man, feeds either directly or at second-hand, on *green plants*. Only plants can use the energy of the sun to build up starch and other

Insects

foods from the oxygen, carbon and hydrogen they get from earth, air and water. Perhaps you have learnt about this process of *photosynthesis* in your science lessons.

Insects live only for a short time, but they have an important place in the whole scheme of life. As we have seen, they quickly adapt to new situations and so we find them living in all kinds of unlikely places, including houses.

Useful books about insects

1. *Insect Natural History* by A. D. Imms (Collins, 1968).
   About all kinds of insects. It has a great many illustrations, some of them beautifully coloured.
2. *Pleasure from Insects* by Michael Tweedie (David & Charles, 1968).
   Mainly about enjoying insects out of doors, but much of it is true for house insects, too. He talks about looking at insects, attracting them, keeping them and photographing them.
3. *Curious Naturalists* by Niko Tinbergen (Country Life, 1958).
   These were naturalists whose curiosity led them to study all kinds of creatures out-of-doors and carry out experiments to find what the creatures would do. Many of these were to find out if insects could learn and if so how.
4. *Insects and Spiders* by Friedlander and Priest (Pitman, 1955).
   Good clear drawings of insects, which will help you to identify those you find.
5. *The Insects* by Peter Farb and others (Time Life International, 1964. Life Nature Library).
   If you have become interested in insects, you will enjoy the pictures in this book, many of them in true-to-life colours. It is interesting to read, too.

Some of the creatures in the food chain.

Spiders and mites   6

How they are different from insects — and from each other

Spiders   What attracts some of them to houses
          Their webs
          How they are useful

Mites   Those that lived in houses in the past

Note   There are a good many questions in this chapter. So have your note-book ready to jot down the answers; then you can check them afterwards.

Many people would call spiders and mites insects, but this is not correct, although they are fairly closely related to each other. What do you think is the most striking thing that insects possess and spiders do not? (Think of how they get about.) And what do most spiders do to catch their prey that insects cannot? Apart from wings and webs, however, if you examine an insect, a spider and a mite, you will notice that an insect has *three* distinct parts to its body — a head, a chest and an abdomen. A spider has only *two* parts — head and chest are joined together, with a separate abdomen. In a mite, the body is in only *one* part.
On its head an insect has a pair of compound eyes and a pair of antennae. (Be sure you know the meanings of these words.) If you examine a spider closely, using a hand-lens or a microscope, you will find that it has six or eight simple eyes, a pair of 'palps' which are very much like antennae, and also a pair of poison fangs. Do you know what it uses these for?
How do you suppose a spider can see with all these eyes, some looking forwards, some sideways and some upwards? Do the 'pictures' from all the eyes form one picture in the spider's brain — or does it attend to the pictures from only some of its eyes at a time? At present we do not know. Are some eyes used for seeing

things far away and some for things close at hand? Probably, but we are not yet sure. Sometimes a spider has some white eyes and some darker ones. Does it use the white eyes in poor light and the darker ones in bright light? This, too, seems likely, but hasn't yet been proved. Try to think of some tests that could be used to help in solving these problems.

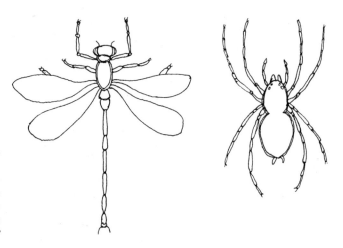

Left, An insect, showing the head with a pair of antennae, the chest with three pairs of legs and two pairs of wings and the segmented abdomen. An insect has three parts to its body.
Right, A spider has two parts to its body, eight legs and usually six or eight eyes. Can you see palps, poison fangs and spinnerets in the diagram?

The insect's chest is in three parts, one behind the other, with a pair of legs on each part and a pair of wings on two of the parts. This means, of course, that there are usually six legs and two pairs of wings. (Flies have only one pair of wings. Can you find the name of the little knobs on stalks that have taken the place of the other pair? What does the fly use them for?) Spiders have *eight* legs on the front part of their bodies and usually four or six spinnerets at the back. The name tells you what spinnerets are used for! Mites are closely related to spiders and have eight short legs.

Neither spiders, mites nor insects have a skeleton inside their bodies like ours; instead, the *out*side of their bodies is hard, rather like a shell. How is this an advantage? Can you see now the reason for having their bodies 'segmented'? (This means they are divided into parts.) How could they move otherwise? Think of all the bones and joints in our bodies!

We have seen that insects normally pass through three stages in their life-cycle — as a larva, a pupa and an adult insect. Spiders

Spiders and mites

and mites, on the other hand, are almost exactly the same when they are born as when they are grown-up, but of course they grow bigger, so why don't they burst their skins? In fact, they *do*. A new skin forms under the old one and then the old skin splits and the spider struggles out and gradually draws its eight legs out of their old covering too. Then it moves away and waits a day or two for the new skin to harden. Incidentally, if you can find a spider's cast-off skin, you will see that it looks just like the spider who once lived in it. Such a skin is often mistaken for the withered corpse of a spider that has died of hunger and thirst.

Spiders

The cobwebs often found in houses and sheds are woven by the *common house spider*. It is rather less than half an inch long and has long legs. Its body is mainly light brown, with black edges in front and dark spots on the abdomen. It does not build a circular orb-web, but a sheet-web, with a little tube at the side where the spider can wait until a fly or other insect lands in the web. It moves about on *top* of its web, and this is often a useful way of distinguishing it from the other spider often found in houses.

You will find it quite easy – and very interesting – to keep one of these spiders yourself. It can go without food and drink for some

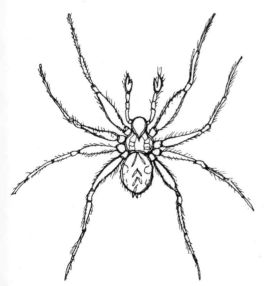

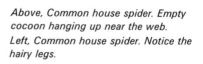

Above, Common house spider. Empty cocoon hanging up near the web.
Left, Common house spider. Notice the hairy legs.

time, so it can safely be left for a time, even if you go away for a short holiday, but it cannot live in a room which has constant central heating. This is too dry for it.

A plastic food-box makes an ideal home for your spider, but any kind of transparent box can be used. Then you need to glue into one corner something like a rolled cylinder of paper for the spider to use as a shelter. You will find that the spider will line this with silk and begin to spin its web quite soon. Wait a few days until it has had time to make a good web, which will probably fill most of the box, and then drop in a fly. Watch carefully to see what happens. Does the spider *see* the fly? You can check by seeing what happens in the dark. How else could the spider know how to find the fly?

This kind of spider takes about a year to become full-grown and it may live for a year or two after that. While it is growing up, it will have several moults, as we have seen, so don't be alarmed if your pet hangs upside-down for a few days and takes no interest in the food you offer it. Just leave it alone and see what happens. An illustrated diary about your spider would be very interesting. The scientific name of this spider is *Tegenaria domestica* and you really need to know this so as to be able to distinguish it from the other kind of house spider, which is called *Pholcus phalangioides* – the Daddy-Long-Legs spider. (Don't confuse this spider with the Daddy-Long-Legs found in the fields.)

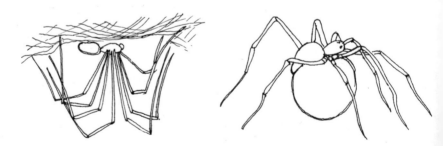

Left, Pholcus hanging upside down. Right, Pholcus, a female with her egg-sac.

Pholcus can often be seen hanging upside-down from its rather untidy-looking criss-cross web, which is usually in corners at the top of walls or on ceilings. Its legs are so very long and thin that it really looks quite comical in this position, as you can see in the sketch. It is mainly pale yellow in colour, with a small rounded 'head' and a long tube-shaped abdomen. You may be able

Spiders and mites

sometimes to see the female spider clutching her round white egg-sac. These spiders feed on mosquitoes, clothes-moths and also in the autumn on several kinds of spiders that come into houses at that time of year.

What do you think attracts spiders to live in houses? It will help you to find the answer if you go around your house noting down where you find spiders. In older houses a thatched roof and a cellar will be the best places to look in. What about a more modern house? Like other creatures, spiders find *shelter* in houses, but also suitable places to spin their webs. Then you might go round again, putting a W in your notebook for each spider or web you find in a wet or damp place and a D for those in dry places, and see how many there are of each. Can you see why spiders usually prefer damp places? There are two main reasons:
1 Spiders die if they become too dry.
2 There are more insects in damp places than in dry ones.

Neither of these two house spiders spins the typical spider's web. This is the work of the garden-spider or diadem-spider. (What is a diadem?) Because of the white cross on her back, many people have regarded this spider as sacred. It may have been from this that there grew up a belief that spiders are lucky. An old rhyme, well-known in America as well as in Britain, says:

> If you wish to live and thrive,
> Let a spider run alive.

If you decide to include the garden with your house, you could make some interesting sketches, or lino-cuts, perhaps, of this spider and how she spins her web. There is a very interesting display at the Natural History Museum in London showing exactly how she does it. (Postcards are on sale there, too.)

You might like to try to build a web yourself, but you would be wise to make it a very big one, for you will find you are not nearly as skilful as a spider! – and of course you will not be able to use such delicate material as spider silk.

We have mentioned earlier one of the chief uses of spiders. They keep down the numbers of insects and in particular of some that spread disease. On the other hand, they provide food for other spiders, for toads and frogs, for birds like starlings and tits and for bats and shrews. Cattle, sheep and horses cropping the grass often eat spiders as well, and of course some are destroyed by larger creatures like ourselves, who crush them as they walk along.

Perhaps partly because they were thought to be lucky, spiders were used in various medicines right up till last century, while in Shakespeare's time and even earlier their webs were used to stop the flow of blood from a wound or cut.

Mites

As their name suggests, mites are tiny creatures. An adult flour-mite, for example, is about 1/50 inch long, so that if you see these mites falling from a sack of grain or flour, they look just like a fine dust.

In the past, there were three kinds of mite often found in houses: flour-mites, cheese-mites and furniture-mites, and also another mite that fed on the flour-mites when there were a lot of them. All were very small, and as you can see from their names, the best way to tell them from each other was by where they lived.

An adult *flour-mite*, of course, has eight legs, but its larva has only six. After a few days the larva moults and emerges with eight legs! After a second moult, one of three things can happen:

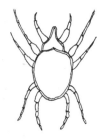

A mite, a small relative of the spider, has only one part to its body.

1 It may become an adult mite.
2 It may pass through a 'resting-stage', in which it has a sucker by which it fastens itself to a mouse, fly or beetle, perhaps, and so gets carried away from its first home. (Do you see the advantage of this? And do you know a *plant* whose seeds are hooked or prickly and get scattered by animals in the same way?)
3 It may have a resting-stage in which its legs almost disappear and its body becomes so light that it can easily be carried away by the wind. (Many plants and trees have very light seeds scattered by the wind too.)

Even a few flour-mites will make grain or flour taste musty, but these mites do not thrive unless the grain or flour is damp.

Cheese-mites look very much like flour-mites and they too like the damp. They do not have a resting-stage in their life-cycle, however. If you do not know about these mites, look for a Stilton cheese in a grocer's shop. You may be able to see in it a small hole which seems to be filled with brown dust. It you could see this dust under a microscope, you would find that it contains a

Spiders and mites

number of live cheese-mites as well as their cast skins and excreta. So if you eat this kind of cheese you may be eating live mites, but they will do you no harm!

Furniture-mites could sometimes be found in hay or straw, or in grain or flour, but sometimes some were blown into the house from outside. In Queen Victoria's time, most houses had a horse-hair sofa in their drawing room or parlour, but very often the sofa was not really stuffed with horse-hair, but with a cheaper substitute called green Algerian fibre. As a rule, the room was seldom used and so it became rather damp and the furniture rather mouldy. This means that there was a tiny fungus called a mould growing on it. The mould grew very well on the fibre inside the sofa and even on the leather covering. The furniture-mites crept inside the stuffing and fed on this mould.

You can now see why mites are seldom found in modern houses. These are drier than houses in the past and only small stores of flour and cheese are kept in them, and these stores are kept dry, so that there is nothing to attract the mites.

Helpful books about spiders

1 *The World of Spiders* by W. S. Bristowe (Collins, 1958).
   You will very likely find this book in your school library. It tells you all about spiders and is beautifully illustrated.
2 *Insects and Spiders* by Friedlander and Priest (Pitman, 1955).
   Very good for helping you to identify spiders. Good, clear drawings of them.

7 Visitors

Summer visitors to the garden

Seasonal visitors to the house itself, some attracted by food and others by dampness

Visitors brought into the house

Human visitors

So far we have been thinking about living creatures that leave their natural homes and settle in a house. There are also a great many more that visit a house from time to time, but do not live there always.

Garden visitors
Visitors are even more noticeable in the garden. If you are thinking of including the garden in your project, you could make a list of summer visitors in four groups: mammals, birds, insects and spiders. You are pretty sure to find there are more insects than any other kind of creature, so you will need to put them into these four groups: flies; beetles; butterflies and moths; bees, wasps and ants. The last group all have a tiny waist.
A needlework picture showing these creatures, with a background of trees and plants, would look well and be fun to make. Or you could build up your picture using shapes cut from paper or material.
Now and then, some of these creatures may get into the house by accident, but they do not stay long.

Visitors to the house itself
There are two main groups of these: 1. Toads, slugs, snails and mosquitos; and 2. Various flies.
Can you see what it is that all the first group like? So you

Visitors

would expect to find them in some damp part of the house, such as the cellar. This means, of course, that many more of them visited houses in the past than now. At which time of the year will you look for them? Why do they come then?

Toads

When you were younger you probably went 'pond-dipping' and found out all about frogs and their eggs and tadpoles. Toads, of course, are closely related to frogs, but you can tell them by their warty skins and the fact that they crawl rather than hop like frogs. So how do you think toads' legs are different from frogs'? Both toads and frogs are amphibious — check this word if necessary. Toads' eggs are found in strings ten to twelve feet long — twice your height — twined round water plants. You could look for eggs and tadpoles in watery places in spring and early summer. The tadpole stage lasts for eight to twelve weeks. After this, the toad lives on land and is often found quite a long way from water. The only toads you are likely to meet in your house will obviously be adults.

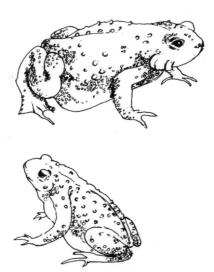

Toads are quite easily tamed, and very soon recognize the person who feeds and looks after them. If you have one for a pet, you will need to keep it in a damp place, but you will notice that it never drinks. Instead, it soaks up water through its skin. Toads will eat almost any small living thing, provided it is moving. They seem to like ants particularly, but they also eat slugs and worms, as well as all kinds of garden pests, so some damp, sheltered spot in the garden will probably be best for your toad, even if you first meet him in the house, which is not very likely nowadays.

Don't be put off, to begin with, if your toad squirts his spare supply of water at you. This is just nervousness and will stop after a week or so if you handle him carefully. By then, he will love to be stroked and petted. You can tell whether your toad is a he or a she, by its size. Males are about $2\frac{1}{2}$ inches long and females about 4 inches.

The warts on a toad's skin are actually glands, which give out a milky juice which is poisonous and tastes horrid to any creature who bites the toad. So don't let your dog near your toad. This juice is not poisonous to human beings and it won't cause warts on your skin, as some people think.

You will find it particularly interesting to watch your toad when it is feeding and also if it is in danger. When an insect flies past near your toad's mouth, you will need to watch carefully to see just how the toad catches it. At first, all you will see will most likely be a flick of the toad's long sticky tongue and the insect is gone. If you look more closely, you will discover that the toad's tongue is hinged at the front of its mouth, instead of at the back like ours. So it is a bit like one of those paper toys that you can roll up and then flick out at your friends.

I doubt if you will discover, or even be able to guess, just what happens to the insect inside the toad's mouth. Of course, it wriggles about, but the toad grips it with its *eyeballs*! I don't think you would guess, either, that the toad can pull its eyeballs right down into its mouth and squash its food with them!

Like many other creatures, a toad 'freezes' if it senses danger. Actually it can alter the colour of its skin slightly, but in any case the skin looks very much like earth, so that when the toad freezes, it is much the same colour as its surroundings and is not at all likely to be noticed by its enemy.

If a toad is really frightened — by a snake, for example — it breathes very deeply and so makes its body swell up and look much bigger than it really is. At the same time, it stretches out its legs as far as possible and makes them quite stiff. This raises its body off the

Visitors

ground and makes it look bigger still. Perhaps the snake will think the toad too large to swallow, even if it isn't really.

Toads often stay in the same neighbourhood for a long time. Some have been taken as far as a mile away and turned up again in their original home. No one yet knows how they find their way back. Perhaps yours will be like the one that lived for thirty-six years underneath the front door-step of a house whose owner fed it regularly!

Slugs and snails

These are molluscs, so that they are closely related to oysters, but I doubt if any oysters will appear in your house. Snails of course carry their houses on their backs; slugs have no shell but make up for it by producing huge amounts of mucus — they leave a trail of slime behind them wherever they go. The best time to find snails and slugs is after heavy rain. Do you see why? If you find them in your house and don't want to keep them there, put them out of doors in a damp spot. But examine them first. You will soon see the purpose of the horns on the head and it will be interesting to watch how these creatures move about. They are also called 'gastropods', which means stomach-foot — they walk on their stomachs. Snail-shells are not by any means all alike. You could make sketches and notes and identify your creatures in a museum or from books.

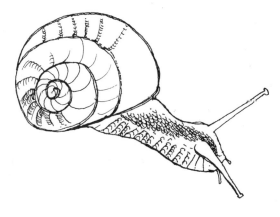

Garden snail. How is this different from a slug?

Visitors

Two mosquitoes resting—Anopheles at the top and Culex below.

Mosquitoes

There are a great many kinds of these insects in Britain, but only two are likely to be found in winter in cellars, out-houses, attics, lofts or similar parts of the home.

If you have ever been bitten by a mosquito, you probably didn't notice anything until it was all over, though if you were in bed at night you were perhaps warned by a high-pitched piping sound. This is made by the insect's wings. The mosquito settles on one's skin so lightly that she is not felt. She feels around until she finds a nice soft spot, then she pierces the skin and sucks the blood. You will notice that I have said 'she', because male mosquitoes do not suck blood. Actually, the commonest mosquito in Britain, called *Culex pipiens*, very rarely sucks human blood but seems to prefer that of birds. If you can look at a female mosquito under a microscope, you will see that, instead of two jaws, it has six 'stylets' for piercing the skin. Each one looks like a very fine, sharp needle.

The other common kind of mosquito you have very likely heard about because it is the one that causes malaria. Did you know that in some wet parts of England quite a lot of people suffered from malaria up to about sixty years ago? It was called 'ague'. I feel sure you have heard how white people could not live in many parts of Africa because of this disease. Now the mosquito causing it has been discovered and the disease almost wiped out. This has been accomplished not only by finding ways of preventing the mosquito from breeding, but also by the discovery of drugs which prevent people from developing the disease if they are bitten.

This mosquito is called *Anopheles*. You can see in the drawing how to tell it from other mosquitoes by the position it rests in. You will see from the other drawing that the egg of the *Anopheles* mosquito has an air-chamber on each side. Why?

a b

Mosquito eggs: (a) Anopheles egg with air-chambers (b) The tiny floating raft formed hundreds of Culex eggs.

Culex mosquitoes lay a huge number of eggs at once, often about three hundred, and all of these form a tiny brownish raft less than $\frac{1}{4}$ inch long. Perhaps you have seen them floating on stagnant

Visitors

water. Another curious thing about these eggs is that they are heavier than water but do not become wet, so the whole raft floats because of the air in the tiny spaces between the eggs.

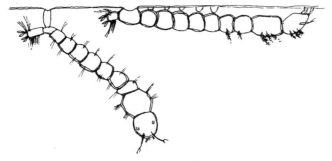

Mosquito larvae: Culex (on the left) and Anopheles. Both kinds of larvae use a 'feeding brush'. Can you see how they breathe?

The larvae of both *Anopheles* and *Culex* mosquitoes move through the water tail first and feed by twirling their 'feeding brush' very fast backwards and forwards. This sweeps tiny particles of food into the larva's mouth. You will see from the drawing that they have a tiny opening for breathing, near their tails. You are likely to find *Culex* larvae in water-butts and puddles but *Anopheles* larvae are more likely to be in slowly flowing water with weeds on it.

So, as with toads, it is only the adult mosquito you will find in the house — or perhaps it will find you!

Flies

The other insects that often visit a house are flies: house-flies, stable-flies, blow-flies and cluster-flies. What do you think attracts *them* to the house? So where would you expect to find them? Some years ago there was a popular song called, 'Where do flies go in the winter-time?' The song did not give an answer because no one knew then. It is now thought that no grown-up flies live through the winter, but their eggs are laid and it is the new generation that hatches out as maggots in the spring and appears fully-grown in the summer.

House-flies

These flies will eat anything. In the nineteenth century John Ruskin described the house-fly 'feasting at his will with rich variety of feast from the heaped sweets in the grocer's window to those of the butcher's backyard'. You might draw a picture of such a sweetshop window buzzing with flies in the days before sweets were wrapped and sold in sealed bags and packets. Another picture might show the flies swarming around the offal in the butcher's yard before our modern very strict laws were made about this.

The housefly carries millions of disease germs in and on its body.

You could go to the Health Department of your Town Hall, or Council Offices, and ask exactly what rules have been made about this in your district. It would be interesting, too, to find out what steps the authorities take to see that these rules are kept.

Your pictures would be even better if your match them with others showing the much cleaner conditions today, with *no* flies. Perhaps you can also think up witty titles for your pictures.

Although house-flies are much scarcer now, those that do exist are still a great danger to health, as I expect you know. One of their favourite foods is human excrement with all the disease germs it contains. These germs collect on the fly's feet and around its mouth. Then if the fly finds any human food — sugar, jam, milk, bread, cheese, meat, fruit — left uncovered, it eats this and of course spreads the germs.

Visitors

In summer these germs multiply very quickly, especially in foods like milk, and from drinking this milk people may get diseases like cholera, typhoid and dysentery. In Britain the spread of summer diarrhoea among babies is believed to be caused by flies infecting their milk.

It is interesting to notice, by the way, that flies 'eat' by sucking or sometimes by piercing and then sucking; they cannot eat solid food. Flies like to lay their eggs on fresh horse-manure, but they will use other decaying matter, including household rubbish.

You might try now to see how complete a list you can make of ways to prevent the spread of disease by flies. This would be better still as a picture chart.

Other flies

From August to October swarms of very small flies, not more than a sixth of an inch long, sometimes appear on window-panes and ceilings of houses. It is curious that they seem to visit the same houses year after year. These are not small house-flies growing up, because of course once a fly has its wings it is grown-up and doesn't grow any bigger. These little flies have no easy English name.

Another kind of fly is one that bites people in the late summer. This again is not a house-fly, though it looks like one; it is a *stable-fly*.

The *blow-fly*, often called a blue-bottle, is another fly often seen in houses in the summer. Can you find out why it has these two names? It is larger than the house-fly and often buzzes loudly as it flies. Although, like the house-fly, it normally only feeds on liquids, it can also lift its 'lips' and use the very fine 'denticles' in its mouth to saw off pieces from a lump of sugar, for example. Blow-flies generally lay their eggs on meat and they develop into the maggots you have perhaps seen on a joint of ham or other meat. On the other hand, blow-flies are very useful scavengers. What does this mean that they do?

Another fly, very much like the house-fly but rather bigger, appears in large numbers clustering together to sleep through the winter. These *cluster-flies* are likely to be found in disused rooms, lofts and so forth and you might find them behind boards and curtains that are not often disturbed.

Other visitors to the house

A great many people in the past liked to have flowers and pot-plants in their houses, as we still do. Various creatures may be carried into the house with the plants: earthworms, earwigs, greenfly, blossom beetles, and lacewing flies, which feed on the greenfly. Can you work out which live on the flowers, which on the stems and leaves, and which in the soil? You could make a

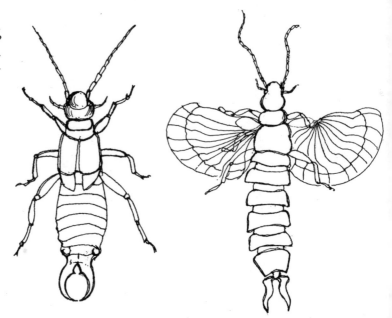

Left, Common earwig. Can you see its wings? Right, Common earwig, flying.

sketch of the plant with these creatures on it in their right places. In the past, too, doctors sometimes carried leeches into houses, in order to 'bleed' their patients. As you very likely know, bleeding was for a long time one of the chief ways of treating illness. How much good do you think it did?

Quite a lot of creatures were carried into the house with the firewood, too. If you still sometimes burn logs, you will know some of these creatures that normally live underneath bark or in the cracks of it. Among them are woodlice, millipedes, centipedes, slugs, snails, springtails and beetles. Here are drawings of some of them, which you could study in order to see how they are different from each other. You could use them, too, in making an illustration of a log bringing living creatures into the house. You might show the log as a living thing walking along.

A medicinal leech

Visitors

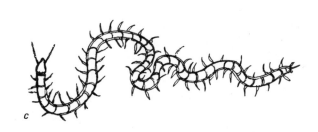

Creatures carried into the house with the firewood: (a) woodlouse (b) millipede (c) centipede (d) springtail.

If you think about the food of these various visitors, and about the creatures that prey on them, you will be able to work out for yourself how these visitors would alter the lives of the 'residents' in the house. This would look well, shown as a number of food-chains in various colours. (See page 39.)

If you like, you can also think about human visitors to the house and the special arrangements that might be made for their visits, such as special meals to be cooked, cleaning and polishing. What else can you think of and draw? You could bring out the fact that, with easier travel, visitors are far more usual nowadays and it is much easier to make them comfortable.

To show all the visitors, you could sketch the house with human visitors being met at the front door, while toads, snails and slugs crawl up to the back door perhaps and mosquitoes and flies fly in at the windows, and so on. You could think of an interesting way of doing this, perhaps by covering the outline shapes with glue and then dropping small coloured beads or different coloured seeds onto them.

Other books to read
1 *Amphibians and Pond-dwellers* by Maxwell Knight (Museum Press), a book in the series, 'Instructions to young Naturalists', which may tempt you to have a pond in the garden of 'your' house.
2 *The Young Specialist looks at Molluscs* by H. Janus (Burke, 1965).
3 You can find out more about *flies* and *mosquitoes* from the books mentioned at the end of Chapter 5, on page 40, especially the first two.

Parasites 8
Creatures that live on others, or in them, and feed on them

Different ways in which one creature may live on another

Various kinds of creatures in a house that may have parasites living on them

These parasites are usually fleas, lice, bugs, worms, mites or flies

On the whole, of course, a parasite is a small creature living on one a good deal larger, but you may have heard the rhyme:

> Great fleas have little fleas
> Upon their backs to bite 'em;
> Little fleas have lesser fleas,
> And so ad infinitum.

This means, of course, that small creatures living as parasites on other creatures may themselves act as 'hosts' for yet smaller ones.

When we call a person a parasite, we mean a good-for-nothing who sponges on other people for his needs, but in Nature, as we have seen, different kinds of living creatures get their food either at first-hand from plants or at second-hand from some other kind of creature. Parasites get theirs, as it were, at third-hand. So living as a parasite is just another way of life. Creatures that feed on plants are herbivores; those that feed on flesh are carnivores; those that feed on either are omnivores.

It would be interesting to write down the pros and cons (advantages and disadvantages) of each of these ways of life and of living as a parasite. Which kind of creature gets its food most easily? Which kind has the most reliable food-supply? Which runs least risk when feeding? Think about how the creature has become adapted to its way of life and whether in doing this it has exposed itself to other risks. If you leave plenty of space, you will be able to fill in your chart as you find out more about each of these ways of living.

I expect you have already thought of the greatest advantage a parasite has — it drinks the blood of its host and so, once it has found a suitable host, it can have a meal whenever it likes, whereas most living creatures have to spend nearly all their time looking for enough food to keep them alive.

Because of its small size, a parasite does not usually do much harm to its host. The death of its host is, of course, the greatest threat to the life of a parasite. If the host dies, so does the parasite, unless it can immediately find another host of the same kind.

Parasites do not all live in the same way. Some live outside their hosts and some inside. Can you find examples? The sheep liverfluke lives part of its life-cycle in the sheep's liver and part in a special kind of snail. Try to find out how it gets from one host to the other. Most parasites live in one special kind of host only, and very often in one particular place in that host. You could make another list to show the pros and cons of living inside or outside your host, and of having one host or more.

Parasites on living creatures in a house

If a creature living outside a house has parasites, it will carry them with it when it goes to live in the house — or of course there may be small creatures in the house, mites for example, which may take the chance of an easier way of life if a suitable host appears. When we come to think again about the various living creatures in a house — people and their pets, other mammals, birds, insects, spiders and mites, and even visitors like flies — we shall find that they may all have parasites and many of them have more than one kind.

In most houses today there are not many parasites living on the *people*, but in the past fleas, lice, bugs and at times worms and mites were a great nuisance and even a danger. Here are some reasons why there are fewer parasites now:

1 People wash more often and more thoroughly.
2 Hair is brushed and combed and washed regularly.
3 Disinfectants can be used.
4 Houses are cleaned oftener and more thoroughly.
5 Food is kept much cleaner.
6 It is easier to cook food more thoroughly.

Parasites

You could write out this list, or one of your own, and add any other reasons you can think of. Leave a good space between each line. Then under each reason you could sketch the parasites you think became fewer for that reason.

The housewife in earlier times had to make her own ointments to use against fleas and lice and she had to see that she always had plenty of these ointments. It was generally easier to get rid of lice than fleas. Can you see why? Both live on the human body but, while lice spend most of their lives on the body, fleas only live there when they are adult. They lay their eggs in cracks in the floor and the larvae develop there. Even as adults, fleas can hide among the bed-clothes or in similar places.

There are now two kinds of louse found on people — the head-louse and the body-louse. Which do you think came first? Before people wore clothes, the lice lived among their hair for protection and fastened their eggs to it, too. The body-louse only came when people started to wear clothes. If the hair was kept well brushed and combed, clothes washed and a suitable ointment used, lice could be got rid of, but adult fleas which were killed would quite likely be replaced by others which had grown up elsewhere.

Fleas are the parasites most often found on pets and other mammals, while birds and their nests provide homes for fleas, lice and mites. Insects and spiders sometimes have insect-parasites.

Fleas

The different kinds of fleas that may now live on people, dogs, cats, rats, mice, bats, or birds are all descended from the same kind of creature. We know that there were fleas very much like modern ones living as parasites fifty million years ago, because one has been found preserved as a fossil in a piece of amber of that date. Perhaps you are wondering how we know it was a parasite. You will find the answer if you think carefully as you read on.

The ancestor of that flea, even more millions of years ago, seems to have been a fly with wings which lived by scavenging among the rubbish in the underground burrow of some mammal and happened to creep into its fur. It soon discovered that it could get an easier and more nourishing meal by sucking its host's blood.

Since then fleas have changed a great deal in order to fit in better with this new way of life. It is easy to see that wings would be a nuisance to a flea living among an animal's fur or hair, so in time fleas no longer had wings. As they could not then escape by flying away, they gradually developed legs with stronger and stronger muscles, which enabled them to take longer and longer leaps. A rat flea, which is not much bigger than a pin's head, can jump a distance of over a foot. The record jump for a human being is less than thirty feet. Yet a human being is very much more than thirty times the size of a flea. Human fleas can jump about seven inches.

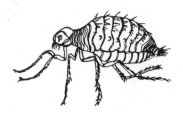

A flea has no wings but very strong legs.

It is possible for a flea to go for a long time without food, but it very quickly realizes when a meal comes within reach. People coming into a room which has been empty for a long time may find themselves attacked by fleas almost at once. In the same way, an adult flea may rest for a long time in its cocoon, but as soon as an animal returns to its burrow (or a man walks across the floor) the flea will hatch out, hop onto the host and start to feed.

So its strong legs help a flea to escape from its host or from anyone who is trying to catch it, but their chief use is to help it to leap onto a host and fasten itself there. The flea jumps too quickly for our eyes to keep up with it, but it has been shown by photographs that the flea very often lands facing the way it has come, so that it must twist over as it jumps. It has even been possible to arrange for a flea to take its own photograph in mid-air, and this photograph shows the flea upside-down with one of its three pairs of legs sticking straight up in the air. You will see that all its legs are covered with stiff bristles and end in strong claws, and I expect you have guessed that the flea uses these to fasten itself to the host's fur or hair as it lands.

Flea bites cause a good deal of irritation, so the host is likely to scratch itself or try to get rid of the flea in some other way. What

Parasites

kind of outer covering would be most useful to the flea as a protection against this?

Here is a drawing made in Germany in 1739 of a woman wearing a flea-trap round her neck and also a larger drawing to show what the trap itself was like. In old books it is always women who are

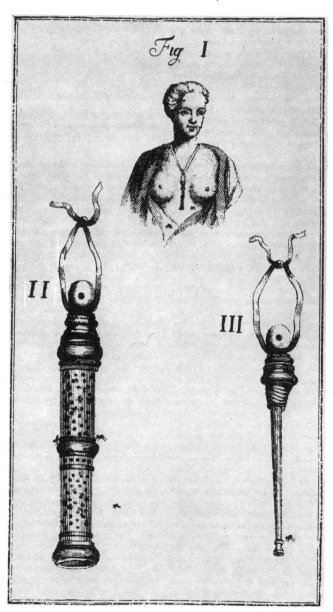

Some eighteenth-century flea traps.

wearing these traps because women were supposed to be bitten by fleas more often than men were. People at the time thought this was because women's skins were more delicate than men's, but now that we know about the chemical messengers in our blood called hormones, it seems more likely that the fleas were attracted by the kind of hormones that women have and men do not.

Rabbit fleas

A pet rabbit may sometimes have fleas on it, but there were many more fleas on wild rabbits. It was these fleas that caused the death of millions of rabbits from myxomatosis a few years ago. The life-cycle of these fleas is linked in an astonishing way with that of the rabbits they live on.

Once they find themselves on a rabbit, the fleas make their way to its ears and fasten themselves there by their mouths. The very curious fact about these fleas is that they can only breed on a rabbit that is pregnant. Ten days before the young rabbits are born, the eggs of the female fleas begin to develop, so that the young fleas are ready to be born at about the same time as the young rabbits. Then the fleas move from the rabbit's ears to her face. While the mother rabbit is looking after her young, the fleas hop onto them and start to feed. After feeding on the young rabbits for a while, the fleas can mate and the females can lay their eggs. After laying eggs in the rabbit's nest for about twelve days, the fleas suddenly go back to the mother rabbit and the whole cycle can start again. Biologists have proved that it is a special hormone in the blood of the mother rabbit that sets off the breeding of the fleas.

Other fleas

Dogs and cats, and even mice, sometimes have their own kinds of fleas, too. If you examine a dog's ear very gently, you will find a little fold of skin rather low down towards the back of the ear. This is a nice warm place for fleas to hide and is known as Henry's pocket. If you know a vet, you can ask him or her why it is called this. Some of the fleas may be dog fleas, but others may be hedgehog fleas that your dog has picked up in the fields, probably from a dead hedgehog. Both dogs and cats can get tapeworms inside them from swallowing fleas, so it is important

Parasites

to get rid of fleas at once. A vet will tell you the best way.
Curiously, fleas do not bite horses because they do not like their smell. Grooms and other people who have a lot to do with horses get the smell from them and so do not get bitten either, so long as they do not wash too much!

Bugs

Another parasite that used to be common in England but is seldom found in houses today is the bed-bug. To begin with, so far as we can tell, this bug lived on bats and no doubt spread to people when they also lived in caves. The first of these bugs in England came from countries near the Mediterranean Sea about three hundred years ago. How did it get here, do you think? When did we start to trade with the Mediterranean countries? See if you can find out the names of some of the famous explorers and traders of that time. What sort of cargoes did they carry in which bugs could hide?

Bed-bugs.

If you examine one of these bugs, or a picture of one, you will notice how flat it is. So it is not hard to think of the kind of place it would hide in, in a house. People found that the best way to get rid of these bugs was to paint all the cracks in the beds and everywhere else in the room with a feather dipped in turpentine. Sometimes bed-bugs were killed by another kind of bug called an assassin bug or by very cold weather. Pigeon-bugs and swallow-bugs from the house-martins' nests sometimes got inside the house, too.

Parasitoids

These are the kind of parasites usually found on insects and spiders, so that the host and the parasite are almost the same size. These creatures, which are usually small flies, are parasites when they are larvae but live independent lives when they are adults. A very great difference between parasitoids and true parasites is that parasitoids nearly always kill their hosts just before they themselves are fully grown and will not need the host any longer. In an earlier chapter we have seen that insects multiply very quickly indeed; if it were not for the parasitoids that kill some of them, there would soon be far too many insects in the world.

Diseases carried by parasites

If too many parasites live on one host, they may make it so weak that it dies, but this does not often happen. Even today, however, some parasites can kill their hosts because of the disease germs they carry. This was much more usual in the past. You could make a list of discoveries and inventions in hygiene and medicine that have helped to wipe out these parasites. It would be interesting, too, to find out something about the people who made the discoveries. In the past, human lice carried typhus fever, mosquitoes carried malaria, while tiny mites were the carriers of scabies in people and of mange in animals. Nowadays these illnesses are very unusual in England.

The fleas that lived on the black rats were the most dangerous to people because they carried the 'plague'. Both the rats and the fleas caught the disease but the rats died of it sooner than the fleas. This particular flea can live either on a rat or a mouse or a person, so when its rat-host died, the flea would be quite likely to bite people and so spread the plague. I expect you have heard about the 'Black Death', which spread across Europe about six hundred years ago and killed a quarter of all the people living there. Of course, even more fleas died than rats or people, but this is an interesting case of the very great difference a tiny creature can make in the lives of others, even if they are much bigger.

Books

1 *Cine-Biology* by J. V. Durden, Mary Field and F. Percy Smith (Pelican Books, 1942).
  Chapter 7 of this book tells about the sheep liver-fluke and there are photographs of this and many other small creatures.
2 Most parasites are *insects* and you can find out more about them from the books on page 40.

Conclusion: very important points to be illustrated or mentioned in your project

1 The various communites of living creatures in a house.
2 The three main needs of all these creatures — food, shelter and a mate.
3 How these needs can be met in a house.
4 How the different communities affect each other.
5 The 'economy' of Nature; an animal is *either* eaten by another creature *or* is killed by germs. The creatures themselves die, but nothing is really wasted.
6 Only green plants can use the substances found in air, soil and water to make the foods needed by plants and animals. Animals cannot do this for themselves. So all the creatures living in a house depend in the end on the green plants and could not live without them.

Some more ideas about things to do

1 Make a series of models or pictures of the chief creatures living in a house. (You can use any of the ways already suggested or of course any others you can think of.) Try to make each creature of about the right size compared with the others. You might make your pictures into a folding frieze.
2 You could decorate your pages, or some of them, with cut-outs of the creatures you are writing about. The cut-outs could be in paper or material and either black or coloured.
3 If you try to use an ordinary small camera like a Brownie for taking snaps of living creatures, you will find it focuses too far away, but you can make it work by buying a cheap pair of spectacles and using one of the lenses. Remove the lens carefully and fasten it securely in front of the lens of your camera. You will see that there is a number on the lens. If it is a number 5 lens, for example, you will be able to take a snap of an insect only 8 inches away from the lens. Be sure the insect is in bright light, or use a flash bulb. Don't be disappointed if your first pictures are not very good. They will soon improve and you will enjoy making your own illustrations in this way. For larger creatures you will need to focus your camera further away:

To focus 10 inches away, use a number 4 lens
$\qquad$ 13 $\qquad$ 3
$\qquad$ $19\frac{1}{2}$ $\qquad$ 2
$\qquad$ 39 $\qquad$ 1

You can find photographs of a boy taking pictures like this and of the kind of picture he can take in a book called *Strange Ways of the Plant and Insect World* by Ross Hutchins. It was published in 1960 by Burke and your library will very likely have a copy.
4 If you can visit the Natural History Museum, South Kensington, London, you can see all kinds of living creatures. The Bird Gallery and the new Insect Gallery are particularly interesting.

How to keep living creatures in order to watch them carefully

Before you do this, you should read about the creature and answer these questions about it:

1 What food does it need?

2 What kind of box or container can it be kept in? How large should this be and what is needed inside it?

3 Will the creature be interesting to watch?

4 Has it a nasty smell?

5 Am I prepared to clean out the cage properly and regularly?

A very useful book for this is *Practical Fieldwork for the Young Naturalist* by Joy O. I. Spoczynska, published in 1967 by Frederick Muller. It has a section on keeping living creatures for study and also gives useful advice about handling living creatures and about cameras and other apparatus you may find useful.

Acknowledgments

The author and publishers would like to thank those listed below for permission to reproduce illustrations:
The Times, for the photograph of Lower Brockhampton, Herefordshire; Sdeuard C. Bisserot, Eric J. Hosking, Mrs N. S. Heriot, John Markham, Leonard Taylor, M. W. F. Tweedie, The Natural History Museum.
Thanks are due to the following for permission to reproduce copyright material:
Methuen & Co. Ltd for 'Milk for the Cat' by Harold Monro from *An Anthology of Modern Verse, 1920-1940*; Macmillan & Co. Ltd and Mrs Hodgson for 'Stupidity Street' from *Collected Poems* by Ralph Hodgson.